Angela Matilde da Silva Alves
Cyro Rêgo Cabral junior
Elizabeth Sampaio

Determination of fungal contaminants in refrigerated milk from AL.

Angela Matilde da Silva Alves
Cyro Rêgo Cabral junior
Elizabeth Sampaio

Determination of fungal contaminants in refrigerated milk from AL.

Characterization of fungal contaminants in milk from expansion tanks produced in the state of Alagoas.

Imprint

Any brand names and product names mentioned in this book are subject to trademark, brand or patent protection and are trademarks or registered trademarks of their respective holders. The use of brand names, product names, common names, trade names, product descriptions etc. even without a particular marking in this work is in no way to be construed to mean that such names may be regarded as unrestricted in respect of trademark and brand protection legislation and could thus be used by anyone.

Cover image: www.ingimage.com

This book is a translation from the original published under ISBN 978-620-2-18416-8.

Publisher:
Sciencia Scripts
is a trademark of
Dodo Books Indian Ocean Ltd. and OmniScriptum S.R.L publishing group

120 High Road, East Finchley, London, N2 9ED, United Kingdom
Str. Armeneasca 28/1, office 1, Chisinau MD-2012, Republic of Moldova, Europe
Printed at: see last page
ISBN: 978-620-7-58375-1

ACKNOWLEDGMENTS

To JEHOVAH God for everything he has done for me and for supporting me in difficult times, giving me the inner strength to overcome difficulties and providing for all my needs.

To my advisors, Professors Cyro Rêgo and Elizabeth Sampaio, for their guidance and trust.

To my family, whom I love very much, for their affection, patience and encouragement. Especially my mother, who has always been an example of strength and overcoming for me, as well as my beloved husband Alex, who was by my side at all times. Not forgetting my little Maísa, who was with me through all my anxieties from the moment she was born.

To Professor Dr. Ângela Froehlich for her help and friendship, for believing in and contributing to my professional growth and for also being an example of a professional to be followed. Her help was important in carrying out this work.

To the friends who were part of these moments, always helping and encouraging me.

To my friends Hugo, Allana Patrícia, Karla Daniele, Valéria and Genildo, who have always been by my side, giving me strength, support and practical help.

LACEN for their help with reagents.

To my master's colleague who became my personal friend, Quitéria Gomes, for sharing in all the stages of this process, helping me to carry on even when discouragement asked me to stop.

To IFAL in the person of Professor Johnnatan Freitas and Cecília Dantas for their exceptional technical support.

GENERAL SUMMARY

SUMMARY

The aim of this study was to count filamentous fungi and yeasts and characterize the fungal contaminants in milk from expansion tanks produced in the state of Alagoas, and to evaluate the hygiene and sanitary conditions of these tanks. Previously sterilized glass containers were used to collect the milk. Immediately after collection, the containers were placed in isothermal boxes containing recyclable ice for later analysis at the Food Inspection Laboratory of the Federal University of Alagoas. The samples were collected in 23 municipalities from 100% of the community tanks of a cooperative in the state of Alagoas. The samples were collected in January, February and May 2013. At each time, 45 samples were collected, totaling 135 milk samples. The methodology proposed by APHA (American Public Health Association) was used to count and identify filamentous fungi and yeasts. The samples were sown on DRBC medium (Dichloran Rose Bengal Chloramphenicol agar). The dilution used was 10^{-3} . The plates were incubated at room temperature (25°C) for seven days. Of the 135 samples analyzed, only 6 samples (4.44%) showed no contamination by filamentous fungi and yeasts, while 26 samples (19.26%) showed a high level of contamination. More than a third of the 49/135 samples (36.3%) had an average of 3.58x105 CFU/mL of milk. Among the fungi isolated, *Penicillium* spp. 18/135 (13.33%), *Aspergillus* spp. 85/145 (62.96%), *Absidia* sp 32/135 (23.70%), *Fonsecaea* sp *45/135* (33.33%) *and Candida* spp 19/135 (14.07%) stood out, being statistically higher when compared to the other fungi isolated. Due to the high count of filamentous fungi and yeasts, and the variety of their fungal contaminants found in this study, it is suggested that efficient means be adopted to reduce the multiplication of these microorganisms, in order to avoid the problems caused by their metabolites.

Keywords: Milk. community tanks. count. fungal contaminants.

TABLE OF CONTENTS

GENERAL INTRODUCTION

1. GENERAL INTRODUCTION

As a complete food from a nutritional point of view, milk is an ideal substrate for the development of various microorganisms (CALLON et al., 2007; DELAVENNE et al., 2011). Their presence and/or quantity is related to the quality of the raw materials, the manufacturing technology used and the distribution of the product (SPANAMBERG et al., 2008).

Fungi, when present in milk, secrete hydrolytic enzymes such as proteinases, lipases and phospholipases that act on the substrate, altering its physical and chemical characteristics and causing damage to the industry. They can be correlated with the level of hygiene in the milking parlor and the environment, and can also be associated with infectious mastitis. Among fungal contaminants, yeasts are most frequently linked to mammary gland infections in dairy cattle. In dairy farming, mastitis is a major problem for the industry as it reduces the quality and quantity of milk, altering its physical and chemical characteristics and causing economic losses (SPANAMBERG et al., 2008; NORNBERG, TONDO, BRANDELLI, 2009; ZARAGOZA et al., 2011).

Fungi belonging to a variety of genera and species can cause serious damage to human and animal health, such as: allergies, infections, opportunistic mycoses, whether superficial or deep, and skin infections (FLORES, ONOFRE, 2010; SOUZA et al., 2010; CRIADO et al., 2011). Fungi that produce mycotoxins, which are secondary metabolites, cause major economic losses and represent a potential risk to Brazilian agribusiness (QUEIROZ et al., 2009).

The aim of this study was to count filamentous fungi and yeasts and characterize the fungal contaminants in milk from expansion tanks produced in the state of Alagoas, and to evaluate the hygiene and sanitary conditions of these tanks.

2. LITERATURE REVIEW

2.1 THE REGIONS OF ALAGOAS

The state of Alagoas is divided into 102 municipalities in three mesoregions - Sertão, Agreste and Leste Alagoano. The three Mesoregions are divided into thirteen Microregions: four in the Sertão (Serrana do Sertão Alagoano, Santana de Ipanema, Alagoana do Sertão do São Francisco and Batalha), three in the Agreste (Palmeira do Índios, Arapiraca and Traipu), and six in the Leste Alagoano (Serrana dos Quilombos, Mata Alagoana, Litoral Norte Alagoano, Maceió, São Miguel dos Campos and Penedo) (IBGE, 2014).

The state of Alagoas is the sixth largest milk producer in the Northeast, producing a total of 231,367,000 liters of milk in 2010 alone, representing 0.8% of national production (BRASIL, 2011). The state's dairy basin is made up of 17 municipalities, 12 of which are located in the Alagoas hinterland and 5 in the agreste, with approximately 2,500 producers, generating 25,000 direct jobs, and is important for the state's economy (BNB, 2005).

2.2 MILK

Normative Instruction No. 62, of December 29, 2011, defines milk as the product derived from the complete, uninterrupted milking, under appropriate hygiene conditions, of healthy, well-fed and rested cows (BRASIL, 2011).

From a nutritional point of view, milk is a complete food and its consumption is essential for the child's full development, which is why it should be preserved and promoted (WHO, 2009).

As a complete food, milk is easily susceptible to microbial contamination. Some of these agents are fungi, and since milk contaminated by fungi has its physicochemical and organoleptic characteristics altered, dairy product manufacturers are looking for good quality raw materials (SPANAMBERG et al., 2008; NORNBERG, TONDO, BRANDELLI, 2009).

The quality of the raw milk produced is closely related to the multiplication rate of contaminating microorganisms. Several extrinsic factors can influence the quality of the milk, such as: hot weather, rainy season, animal feed, milking environment, the bacteriological quality of the water used, poorly sanitized utensils and equipment, the health of the milkers and the animals. These contribute decisively to the microbiological quality and chemical composition of the milk. Synergistically, important factors such as good practices, animal feed, prior sanitization of the teats and the milking place, which includes teats, cans, milking machine, floor and adequate facilities are decisive for milk production and reducing contamination by deteriorating and pathogenic microorganisms in milk. It should be noted that poorly sanitized or unsanitized cooling or expansion tanks contribute to the growth of the microbial population (YAMAZI et al., 2010; ROMA JÚNIOR et al., 2009; HECK et al., 2009; KORB et al., 2011).

In Brazil, it was only in the 1990s, with the implementation of Normative Instruction no. 51 of 2002, later replaced by Normative Instruction no. 62 of 2011, that the storage of refrigerated raw milk was regulated. In addition to the reduction in operational production costs, there was also a reduction in milk spoilage due to the microorganisms present. However, as the days passed under refrigeration, the milk had technological problems, because the microorganisms continued their activities even under refrigeration. This was mainly due to factors such as poor sanitization of the equipment used to keep the milk refrigerated (expansion tanks) and animals with mastitis. Therefore, hygiene procedures used in the milk production chain are critical to obtaining high quality raw materials (SANTOS et al., 2009; YAMAZI et al., 2010).

With increased access to information, consumers who are concerned about food safety are demanding better quality products from the industry, and this is also being demanded of producers. In order to achieve better results in the microbiological quality of milk, the Ministry of Agriculture, Livestock and Supply (MAPA) implemented national milk quality standards with the publication of Normative Instruction 62 of 2011, which improves the system for collecting refrigerated raw milk and transporting it in bulk (YAMAZI et al., 2010). However, the microorganism count will only be low in refrigerated raw milk if it is produced under suitable hygienic conditions (SANTOS et al., 2009).

Contamination of milk by fungi can cause various problems, including mycotoxins, which are secondary metabolites of fungi. In addition to mycotoxins, there are also other problems caused by milk contaminated by certain species of fungi, which can be allergenic or even deteriorating in nature, which is detrimental to the dairy industry, which suffers a loss in the quality of its products, as well as to consumers, who eat contaminated food (FLORES, ONOFRE, 2010; SOUZA et al., 2010; CRIADO et al., 2011; SIGNORINI et al., 2012; KABAK et al., 2012).

Brazil was considered the sixth largest milk producer in the world (USDA, 2011). According to EMBRAPA (2012), production increased by 4.5% in 2011 alone. Regulatory bodies, concerned about the quality and safety of this food, have been stepping up their inspections of dairies (PEREIRA et al., 2013). Although there has been an increase in production, the instability of the milk market has led small producers to sell informal milk as an alternative to survive in the market. This practice poses a risk to public health due to the diseases transmitted by raw milk (PEREIRA et al., 2013).

2.3 FUNGI

Fungi are microorganisms that are widely present in nature and are found in practically all environments. They are classified as filamentous fungi and yeasts. Although they are classified as fungi, filamentous fungi and yeasts are morphologically different, and their identification is based almost exclusively on their morphology, both macro and microscopically. Macroscopically, fungi can present various morphological types with filamentous colonies, cottony or cottony, powdery and others (filamentous fungi) and yeasts with creamy colonies and the most diverse types of pigments (TANIWAK; SILVA, 2001).

Yeasts are unicellular, oval and generally larger than bacteria. However, some are elongated or spherical. Filamentous fungi are multinucleate organisms that appear as filaments. The body of a filamentous fungus consists of a mycelium and spores. The mycelium is a mass of rigid filaments and the hyphae are made up of chitin, cellulose and glucose. Yeasts reproduce asexually, usually by budding or twinning. Filamentous fungi reproduce by spore dissemination (TANIWAK; SILVA, 2001).

Fungal nutrition occurs by absorption, with the help of enzymes, where organic molecules are broken down into smaller portions that can be transported more easily into the cell. Among the molds or filamentous fungi of great importance for food safety are *Penicillium* spp. and *Aspergillus* spp. and among the yeasts *Candida albicans* (TANIWAK; SILVA, 2001).

Aspergillus sp. is widely distributed in the environment, where about 185 species of *Aspergillus* spp. have been identified and of these, 20 are documented to cause human disease. Exposure to these agents can lead to colonization, causing allergic manifestations or invasive infection depending on adverse host factors. *A. fumigatus* is the species most frequently associated with cutaneous and systemic infections. It is manifested by respiratory symptoms, chest pain, wheezing (KRISHNAN et al., 2008; ALANGADEN, 2011; MARTINO et al., 2009). The species most closely linked to mycotoxin production is *Aspergillus flavus*.

Many species belonging to the *Penicillium* genus have already been used in research for application in biotechnology, the pharmaceutical industry and the manufacture of milk products. However, some species are producers of toxic compounds such as mycotoxins. Some species of *Penicillium are* also responsible for losses in various agricultural products, including citrus fruits (YAHYAZADEH et al., 2008; LEE et al., 2011; GÓMEZ-SANCHIS et al., 2012).

When a fungus has the ability to form mycotoxins, it requires a number of factors favorable to its development, such as: humidity, temperature, the presence of oxygen, time for fungal growth, the constitution of the substrate, genetic characteristics and competition between fungal strains (QUEIROZ et al., 2009). The *Aspergillus, Penicillium* and *Fusarium* species are probably the most significant mycotoxin-producing fungi in the field (FAO, 2008).

Developed countries have already realized that reducing the levels of mycotoxins in food not only reduces the financial burden on health care, but also brings advantages in international trade such as exports to other international markets (SCHWARZER, 2009).

Candida is a yeast that is part of the microbiota of the human body and animals, colonizing the skin and mucosa of the digestive and urinary tracts, mouth and vagina. In this way, candidiasis is most often of endogenous origin, occurring as a consequence of an immune disorder of the host and the virulence factors of these yeasts, which have the ability to colonize, penetrate and invade tissue (BROWN, 2007; HOLLENBACH, 2008). There are currently around 200 species of yeast included in the *Candida* genus, and among those responsible for infections in humans, *Candida albicans* stands out. This has been reported as the most prevalent, followed by *C. parapsilosis, C. glabrata, C. tropicalis* and *C. krusei* (PFALLER, DIEKEMA, 2007; PANIZO et al., 2009). These yeasts are considered the main group of opportunistic fungal pathogens causing nosocomial blood infections in ICUs (KUMAR et al., 2008; KARKOWSKA-KULETA et al., 2009; NEGRI et al., 2010; COLOMBO et al., 2013).

Candidiasis can occur as a result of the disruption of the parasite-host balance, triggered by the impairment of natural defenses such as the immune system (PLAYFORD et al., 2008). In animals, the presence of fungi can be associated with infectious mastitis and identified through the milk. Yeasts are the fungal agents most frequently related to mammary gland infections in dairy cattle (COSTA, SILVA, ROSA, 2008; SPANAMBERG et al., 2008). The main genera involved are *Candida* and *Cryptococcus, as* well as others such as *Geotrichum, Pichia* and *Trichosporon* (SPANAMBERG et al., 2008).

2.4 FUNGI AND PUBLIC HEALTH

In Brazil, there is a scarcity of reports on the occurrence of pathogenic microorganisms, including fungi, and their involvement in outbreaks of diseases related to the consumption of dairy products. Since the 1990s, there has been an increase in the sale of "informal milk", which is a product without any kind of official inspection, not guaranteeing its microbiological quality. The presence of these fungi in milk can represent a potential risk to human health, especially when you take into account that the consumption of raw milk and its derivatives is common in some countries, even though current legislation prohibits the sale of milk and dairy products without prior heat treatment (RUZ-PERES et al., 2010).

Although fungi can promote desirable aspects from a technological point of view in dairy products, the presence of these microorganisms, when undesirable, can be detrimental from an economic and public health point of view, and can be associated with infectious diseases in humans and animals. These microorganisms can be linked to opportunistic mycoses (infections caused by low virulence fungi) which can coexist peacefully with the host. However, when conditions favor their development, they can be pathogenic to humans (RUZ-PERES, et al., 2010).

Mycotoxin contamination of food causes serious public health problems. This is because the toxic effects caused by mycotoxins on the human body range from acute problems to chronic diseases (WILD; GONG, 2010). When it comes to mycotoxins, there is an aggravating factor because it is very difficult to remove them from food. The most effective way of preventing them is by controlling the growth of fungi in food (ERKEKOGLU et al., 2008). For this reason, the production of infant formulas requires strict control using quality raw materials (MAHDAVI et al., 2010).

REFERENCES

ALANGADEN, G.J. Nosocomial Fungal Infections: Epidemiology, Infection Control, and Prevention. **Infectious Disease Clinics of North America**. v.25, p. 201-225, 2011.

BANCO DO NORDESTE DO BRASIL (BNB). **Profile of the States** - Alagoas: Integrated Development Pole - Alagoas Dairy Basin. 2005. Available at: <http://www.bnb.gov.br/content/aplicacao/Investir_no_Nordeste/Perfil_dos_Eshttp://www.bnb.gov.br/content/aplicacao/Investir_no_Nordeste/Perfil_dos_Estado s/gerados/al_apresentacao.asp>. Accessed on: 29/08/2014.

BRAZIL, Brazilian Institute of Geography and Statistics (IBGE). **Livestock production statistics**. Rio de Janeiro, March 2012. Available at: <http://www.ibge.gov.br/home/estatistica/indicadores/agropecuaria/producaoagrop ecuaria/abate-leite-couro-ovos_201104_publ_completa.pdf>. Accessed on: 29/08/2014.

BRAZIL. Ministry of Agriculture, Livestock and Supply (MAPA). normative instruction No. 62, Brasilia, December 29, 2011.

BROWN, J.M.Y. Fungal infections after hematopoietic cell transplantation. Thomas' Hematopoietic cell transplantation, Third Edition Ed. Blackwell Publishing Ltda, p.683, 2007.

CALLON, C. et al. Stability ofmicrobial communities in goatmilk during a lactation year: molecular approaches. **Systematic and Applied Microbiology.** v.30, p. 547-560, 2007.

COLOMBO, A. L . et al. Recommendations for the diagnosis of candidemia in Latin America. **Revista Iberoamericana de Micología**. v. 30. n°. 3, 2013.

COSTA, G.M.; SILVA, N.; ROSA, C. A. Yeast mastitis in dairy cattle in southern Minas Gerais State, Brazil. **Ciência Rural**, v.38, p. 1938-1942, 2008.

CRIADO, P.R. et al. Superficial mycoses and the elements of the immune response. Anais Brasileiros de **Dermatologia.** v.86, n°.4, p.726-731, 2011.

DELAVENNE E. et al. Fungal diversity in cow, goat and ewe milk. International Journal of Food Microbiology. v.151, p.247-251,2011.

EMPRAPA. **Dairy market situation**. Juiz de Fora, v.5, n.41, p.1-41, 2012. Available at :<http://
www.infoteca.cnptia.embrapa.br/handledoc/948946>. Accessed on: April/2014.

ERKEKOGLU, P.; SAHIN, G.; BAYDAR, T. . Special focus on mycotoxin contamination in baby food: Their presence and regulation. **Fabad Journal of Pharmaceutical Sciences**. v.33, p. 51-66, 2008.

FOOD AND AGRICULTURE ORGANIZATION OF THE UNITED STATES. MICOTOXINS.   2008.   Available at   :
<http://www.fao.org/wairdocs/X50120/X5012o01.htm>. Accessed on: April/2014.

FLORES, L.H.; ONOFRE, S.B. Determination of the presence of anemophilic fungi and yeasts in a Health Unit in the city of Francisco Beltrão - PR. **Revista Saúde e Biologia**, v.5, p. 22-26, 2010.

GÓMEZ-SANCHIS, J. et al. Detecting rottenness caused by Penicillium genus fungi in citrus fruits using machine learning techniques. **Expert Systems with Appplications**, New York, v. 39, p. 780-785, 2012.

HECK , J. M. L. et al. Seasonal variation in the Dutch bovine raw milk composition. **Journal Dairy Science**, Champaign, v. 92, n. 10, p. 4745-4755, 2009. Available at: <http://www.mendeley.com/research/>. Accessed on: April/2014.

HOLLENBACH, E. Invasive candidiasis in the ICU: evidence based and on the edge of evidence. **Mycoses.** v. 51, p. 25-45, 2008.

IBGE. **Municipal Agricultural Survey.** Available at: <http://www.sidra.ibge.gov.br>. Accessed in: March/2014.

KABAK, B. Aflatoxin M1 and ochratoxin A in baby formulae in Turkey: occurrence and safety evaluation. **Food Control**, v. 26, n. 1, p.182-187, 2012.

KARKOWSKA-KULETA, J.; RAPALA-KOZIK, M.; KOZIK, A. Fungi pathogenic to humans: molecular bases of virulence of Candida albicans, Cryptococcus neoformans and Aspergillus fumigatus. **Acta Biochimica Polonica**, v.56, p.211 224, 2009.

KORB, A. et al. Human health risks of antibiotic use in the dairy production chain. **Revista de Saúde Pública Santa Catarina**, Florianópolis, v. 4, n. 1, p. 21-36, 2011. Available at: <http://www.esp.saude.sc.gov.br/sistemas/>. Accessed on: April/2014.

KRISHNAN, S.; ELIAS, K.M., PRANATHARTHI, H.C. Aspergillus flavus: an emerging nonfumigatus Aspergillus species of significance. **Mycoses.** v.52, p. 206-222, 2008.

KUMAR, C.P.G. et al. Carriage of Candida species in oral cavities of HIV infected patients in South India. **Mycoses.** v.52, p. 44-54, 2008.

LEE, M. H. et al. Mutations of p-tubulin condon 198 or 200 indicate thiabendazole resistance among isolates of Penicillium digitatum collected from citrus in Taiwan. **International Journal of Food Microbiology**, Amsterdam, v. 150, p. 157-163, 2011.

MAHDAVI, R. et al. Determination of Aflatoxin M1 in Breast Milk Samples in Tabriz-Iran. **Maternal Child Health Journal**, v. 14, p. 141-145, 2010.

MARTINO, R. et al. Lower respiratory tract virus infections increase the risk of invasive aspergillosis after a reduced-intensity allogeneic hematopoietic SCT. **Bone Marrow Transplant**. v.44, p.749-756, 2009.

NEGRI, M. et al. Examination of potential virulence factors of Candida tropicalis clinical isolates from hospitalized patients. **Mycopathologia**. v.169, p.175-182, 2010.

NORNBERG, M. F. B. L.; TONDO, E. C.; BRANDELLI, A. **Psychrotrophic bacteria and proteolytic activity in refrigerated raw milk**. Porto Alegre, 2009. Available at: <http://www. ufrgs.br/actavet/37-2/art825.pdf>. Accessed on: April/2014.

PANIZO, M.M. et al. Candida spp. in vitro susceptibility profile to four antifungal agents. Resistance surveillance study in Venezuelan strains. **Medical mycology**. v.47, p.137-143, 2009.

PEREIRA, et. al. Microbiota mesophilic aerobic contaminant UHT milk. **Revista Instituto Laticínios Cândido Tostes**, Juiz de Fora, v. 68, n°. 394, p. 25-31, Sep/Oct, 2013.

PFALLER, M.A.; DIEKEMA, D.J. Epidemiology of invasive candidiasis: a persistent public health problem. **Revista Microbiologia clinica**. v.20, p.133-163, 2007.

PLAYFORD, E.G.; MARRIOTT, D.; NGUYEN, Q. Candidemia in nonneutropenic critically ill patients: risk factors for non-albicans Candida spp. **Critical Care Medicine**. v.36, p. 2034-2039, 2008.

QUEIROZ, V. A. V. et al. **Good practices and HACCP system in the post-harvest phase of corn**. Sete Lagoas: Embrapa Milho e Sorgo. Technical circular, no. 122, p. 28, 2009.

ROMA JÚNIOR, L. C. et al. Seasonality of protein content and other milk components and their relationship with the quality payment program. **Arquivo Brasileiro de Medicina Veterinária e Zootecnia,** Belo Horizonte, v. 61, n.6, 2009. Available at: <http://www.scielo.br/scielo.php7pid-sci_arttext>. Accessed on: April/2014.

RUZ-PERES M. et al. Resistance of filamentous fungi and yeasts isolated from raw bovine milk to pasteurization and boiling. **Arquivo Brasileiro de Medicina Veterinária e Zootecnia.** v.17, n°.1, p.62-70, mar.; 2010.

SANTOS, P. A. dos et al. **Effect of refrigeration time and temperature on the development of psychrotrophic microorganisms in refrigerated raw milk collected in the macro-region of Goiânia, GO.** Goiânia, 2009. Available at: <http://revistas.ufg.br/index.php/vet/ article/view/3522/6037>. Accessed on: April 2014.

SCHWARZER, K. Harmful effects of mycotoxins on animal, physiology. In: 17 **th Annual ASAIM SEA Feed Technology and Nutrition Workshop**, Hue, Vietnam. 2009.

SIGNORINI, M. L. et al. Exposure assessment of mycotoxins in cow's milk in Argentina. **Food and chemical toxicology**, v. 50, n. 2, p. 250-7, 2012.

SOUZA, A.E.F. et al. Fungal microbiota of public hospitals in the city of Campina Grande - PB. **BIOFAR**, v.4, p.102-116, 2010.

SPANAMBERG A. et. al. **Mycotic mastitis in ruminants caused by yeasts.** Ciência Rural. v.39, no.1, p. 20, 2008.

TANIWAKI, M.; SILVA, N. **Microbiologia**: fungos deteriorantes em alimentos. Campinas: EDITORA, 2001.

UNITED STATES DEPARTMENT OF AGRICULTURE (USDA). **World agricultural supply and demand estimates. 2011**. Available at: <http://www.usda.gov/commodity/latest.pdf. Accessed on: April/2014.

WILD, C. P.; GONG, Y. Y. Mycotoxins and human disease: a largely ignored global health issue. **Oxford Journals Life Sciences & Medicine**, Carcinogenesis, v. 31, p. 71-82, 2010. ISSN 0143-3334

WORLD HEALTH ORGANIZATION. *Dampness and mould: WHO guidelines for indoor air quality*. Denmark: Copenhagen; 2009.

YAHYAZADEH, M. et al. Effect of some essential oils on mycelial growth of Penicillium digitatum Sacc. **Word Journal of Microbiology and Biotechnology**, Oxford, v.24, p. 1445-1450, 2008.

YAMAZI, A. K. et al. **Production practices applied to control microbial contamination in raw milk production**. Viçosa, 2010. Available at: <http://www.seer.ufu.br/index. php/biosciencejournal/article/view/7210/5136>. Accessed on: April/2014.

ZARAGOZA C. S. al. Yeasts isolation from bovine mammary glands under different mastitis status in the Mexican High Plateu. **Revista Iberoamericana de Micología**. México . v.28, n°.2, p.79-82, 2011.

3 RESULTS ARTICLE

Characterization of fungal contaminants in milk from expansion tanks produced in the state of Alagoas

SUMMARY

The aim of this study was to count filamentous fungi and yeasts and characterize the fungal contaminants in milk from expansion tanks produced in the state of Alagoas, and to evaluate the hygiene and sanitary conditions of these tanks. Previously sterilized glass containers were used to collect the milk. Immediately after collection, the containers were placed in isothermal boxes containing recyclable ice for later analysis at the Food Inspection Laboratory of the Federal University of Alagoas. The samples were collected in 23 municipalities from 100% of the community tanks of a cooperative in the state of Alagoas. The samples were collected in January, February and May 2013. At each time, 45 samples were collected, totaling 135 milk samples. The methodology proposed by APHA (American Public Health Association) was used to count and identify filamentous fungi and yeasts. The samples were sown on DRBC medium (Dichloran Rose Bengal Chloramphenicol agar). A dilution of 10 was used^{-3} . The plates were incubated at room temperature (25°C) for seven days. Of the 135 samples analyzed, only 6 samples (4.44%) showed no contamination by filamentous fungi and yeasts, while 26 samples (19.26%) showed a high level of contamination. More than a third of the 49/135 samples (36.3%) had an average of 3.58x105 CFU/mL of milk. Among the fungi isolated, *Penicillium* spp. 18/135 (13.33%), *Aspergillus* spp. 85/145 (62.96%), *Absidia* sp 32/135 (23.70%), *Fonsecaea* sp *45/135* (33.33%) *and Candida* spp 19/135 (14.07%) stood out, being statistically higher when compared to the other fungi isolated. Due to the high count of filamentous fungi and yeasts, and the variety of their fungal contaminants found in this study, it is suggested that efficient means be adopted to reduce the multiplication of these microorganisms, in order to avoid the problems caused by their metabolites.

Keywords: Milk. community tanks. count. fungal contaminants.

3.1 INTRODUCTION

Milk is a product of animal origin with high nutritional value and is considered a noble product, recommended for the diet of children and adults. Its derivatives have equal nutritional value and are sources of income for the different segments of the milk production chain (RIBEIRO, 2008). Because it is such a noble product from a nutritional point of view, it is common for a variety of microorganisms to develop. These include fungi, which are agents capable of producing a wide range of biologically active substances (RUZ-PERES et al., 2010).

Although fungi can promote desirable aspects from a technological point of view in dairy products, the presence of these microorganisms, when undesirable, contributes to the loss of product quality, causing damage from an economic and public health point of view (RUZ-PERES et al., 2010).

One of the problems linked to the presence of fungi is infectious mastitis. Yeasts are the fungal agents most frequently linked to mammary gland infections in dairy cattle (COSTA, SILVA, ROSA, 2008; SPANAMBERG et al., 2008).

The state of Alagoas has a dairy basin in the sertão mesoregion. There, it is common to consume raw milk, as well as making its derivatives, without first heat-treating it to destroy pathogenic microorganisms. This habit is also common in other parts of the state. Therefore, attention should be paid to the presence of microorganisms, including filamentous fungi and yeasts in milk, which can pose a potential risk to human health. The aim of this study was to count filamentous fungi and yeasts and characterize the fungal contaminants in milk from expansion tanks produced in the state of Alagoas, and to evaluate the hygiene and sanitary conditions of these tanks.

3.2 MATERIAL AND METHODS

3.2.1 Location and sample collection

Three collections were carried out in 45 community expansion tanks at three different times in January, February and May 2013, totaling 135 milk samples in 23 municipalities in the state of Alagoas: Batalha (6), Belo Monte (2), Cacimbinhas (2), Craíbas (2), Girau do Ponciano (2), Ibateguara (1), Igreja Nova (6), Jacaré dos Homens (1), Jaramataia (1), Junqueiro

(1) , Mar Vermelho (2), Paulo Jacinto (1), Penedo (4), Piaçabuçu (1), Porto Real

(2) São Sebastião (1), Senador Rui Palmeira (3), Tanque D'arca (1), Teotônio Vilela (1), Traipú (2), Viçosa (1). 500 mL of milk were collected from each sample in previously sterilized glass containers and transported in isothermal boxes containing recyclable ice to the Animal Products Inspection Laboratory at the Federal University of Alagoas. The samples were kept at refrigeration temperature (4°C±2°C) until the analyses were carried out, no more than 24 hours after collection.

3.2.2 Microbiological analysis

25 mL of the sample was transferred to 225 mL of sterile 0.85% saline solution. Sample and diluent were homogenized to obtain an initial dilution of 10<. After homogenization, 1 mL of this dilution was transferred to a tube containing 9 mL of 0.85% saline solution (dilution of 10^{-2}) until the next dilution of 10^{-3} , for plating and subsequent counting of filamentous fungi and yeasts (in duplicate), according to the methodology of the American Public Health Association (APHA, 2001).

To carry out the plating, 0.1 mL of each dilution was inoculated onto previously prepared plates containing Dicloran Rose Bengal Chloramphenicol Agar (DRBC), spreading the inoculum using a Drigalski loop and incubating at 22-25°C/7 days.

Filamentous fungi and yeasts were counted according to the method described in the *Compendium of methods for the microbiological examination of foods* (APHA, 2001).

3.2.3 Identification of fungal contaminants in milk

After growth and counting, the colonies were selected to prepare slides for direct microscopy. For filamentous fungi, the adhesive tape technique was used, where the adhesive part of the tape was pressed onto the colony and placed in contact with the microscope slide containing a drop of 1% methylene blue. To visualize the filamentous fungus under the microscope, 40x objective magnification lenses were used. The Gram stain technique was used to identify the yeasts. For the yeasts, the structures were visualized using an optical microscope with 100x objective lenses, according to LACAZ, et al. (2002). The structures observed were compared with those contained in mycology atlases, both macro- and microscopically, and identified to genus level.

3.2.4 Experimental design

The various categories are represented as frequencies and proportions. Discrete continuous variables are shown as medians.

Comparisons between proportions were made using the "Z" test for two proportions or the chi-square test with Marascuilo's procedure for 3 proportions. Similarly, the number of CFUs was compared using the Mann-Whitney test when only two groups were compared and the Kruskal-Wallis test when 3 groups were compared.

An alpha value of 5% was adopted for all analyses. The analyses were carried out using the statistical package Epi info v.7 (Epi info Inc, Atlanta, IL).

3.3 RESULTS AND DISCUSSION

Taking into account the facilities where the tanks were located, it could be seen that 34/45 tanks were in suitable hygienic conditions. Of these, 11 were in the Agreste de Alagoas mesoregion, 15 in the East of Alagoas and 8 in the Sertão de Alagoas. It was observed that 08/45 tanks were unlined, and that the surroundings of two of the 45 locations where the tanks were installed were totally disorganized and could even attract pests. 02/45 tanks were in a totally open location. It was also observed that 04/45 tanks in the Sertão de Alagoas had inadequate milk storage temperatures, providing satisfactory conditions for the development of fungi. Of the 45 tanks where the milk was collected, 5 were extremely dirty. Environmental conditions are a determining factor in the hygienic and sanitary conditions of food, which explains the high counts of fungi in milk from community expansion tanks.

Another factor to take into account is the heterogeneity of the samples. Community expansion tanks can have a large number of producers depositing their milk, whether or not these producers have any knowledge of food handling. Lack of knowledge in food handling increases the possibility of contamination, resulting in high counts due to a lack of hygiene and sanitary control.

Out of a total of 135 raw milk samples from 45 community expansion tanks located in 23 municipalities that make up the state's milk cooperative, the presence of fungi was observed in 129/135 (95.56%) of the samples analyzed, with filamentous fungi being present in 125/135 (92.59%) of the samples, while yeasts accounted for 19/135 (14.07%).

With the exception of 06/135 (4.44%) of the samples in which no colonies grew, the colonies isolated from the rest of the samples were above 5000 CFU of fungi per mL of milk.

When the CFU counts were compared between the median of the sunny months and the rainy months, there was no significant difference in the number of CFUs between the sunny and rainy periods.

Considering the 135 milk samples from community expansion tanks, the average count in CFU of fungi/mL in more than a third of the 49/135 samples (36.3%) was 3.58×10^5 CFU/mL of milk. It was also observed that 26/135 samples

(19.26%) were highly contaminated, with countless colonies (>250x103 CFU/mL of milk) (Table 1).

Table 1. Average count of filamentous fungi and yeasts in milk samples from community expansion tanks located in 23 municipalities in the state of Alagoas in January, February and May 2013.

Number of Samples	% samples	Filamentous fungi and yeast count (Average in CFU/g)
06	4,44	0
07	5,18	$5x\ 10^3$
37	27,41	$3,64x\ 10^4$
49	36,30	$3,58x\ 10^5$
10	7,41	$2x\ 10^6$
26	19,26	Uncountable
TOTAL 135	100,00	

Source: Author, 2014

When comparing the medians between the quantities of fungal colony-forming units/mL of raw milk collected from expansion tanks in the mesoregions of Alagoas, it can be seen that there was no significant difference between the sertão mesoregion and the east and agreste mesoregions, not even when the sample was analyzed in a rainy month (May) (Table 2).

Table 2. Comparison of the median CFU of milk samples from community expansion tanks located in the three different mesoregions of the state of Alagoas in January, February and May 2013, stratified by month of collection.

Month	Mesoregion	"n"	Median	P-value[1]	Dunn[2]
January	Agreste	12	32500	0,562	There were no differences
	East	17	10000		
	Hinterland	8	112500		
February	Agreste	13	20000	0,596	There were no differences
	East	17	70000		
	Hinterland	8	385000		
May	Agreste	13	155000	0,393	There were no differences
	East	10	135000		
	Hinterland	10	302500		

[1] P-value referring to the Kruskal-Wallis test.
[2] Dunn's post-hoc test indicates between which groups there is a significant difference (if any). Source: Author, 2014.

With regard to the number of Colony Forming Units of fungi, where growth was on a large scale, making it impossible to count them even though they were collected at different times and in different seasons (sunny months - January and February; and rainy months - May), there was no significant difference in the frequency of samples with the collection period. Although, in the Sertão mesoregion, the fungal count in the first two moments was significantly higher than in the other mesoregions. In May, the East mesoregion, where counting was impossible, had a higher number of samples than the other mesoregions (Table 3).

Table 3. Comparison between the frequencies of milk samples from community expansion tanks in the 23 municipalities in the state of Alagoas with uncountable CFU between the 3 different mesoregions, stratified by month of collection (January, February and May 2013).

Month	Mesoregion	Frequency (%)	P-value[1]	Mar ascuilo[2]
January	Agreste	3/15 (20%)	0,023	Between East and Hinterland
	East	0/17 (0%)		
	Hinterland	5/13 (38,4%)		
February	Agreste	2/15 (13,3%)	0,015	Between East and Hinterland
	East	0/17 (0%)		
	Hinterland	5/13 (38,4%)		
May	Agreste	2/15 (13,3%)	0,194	There were no differences
	East	7/17 (41,1%)		
	Hinterland	3/13 (23,07%)		

[1] P-value referring to the chi-squared test for the 3 proportions of each month.
[2]The Marascuilo procedure indicates between which proportions the significant difference lies (if any).
Source: Author, 2014.

The data obtained in the study made it possible to observe the wide variety of fungal genera isolated from the milk samples analyzed, which totaled 18 genera, including yeasts and filamentous fungi (Table 4).

The frequencies of isolations of 18/135 (13.33%) *Penicillium* spp. (Figure 1 and *2*), 32/135 (23.70%) *Absidia* sp., 85/135 (62.96%) *Aspergillus* spp. *(Figure 1 and 2)*, *45/135 (33.33%) Fonsecaea* sp. (Figure 1 and 2) and 19/135 (14.07%) *Candida* sp. (Figure 2), were statistically higher when compared to the other fungi isolated (Table 4).

Table 4 Isolation frequency of fungi in milk samples from community expansion tanks located in 23 municipalities in the state of Alagoas in January, February and May 2013.

Microorganism identified	Number of samples	%
Absidia sp.	32	23,70
Aspergillus spp.	85	62,96
Actinomadura sp.	02	1,48
Alternaria sp.	09	6,67
Blastomyces sp.	12	8,89
Candida spp.	19	14,07
Coccidioidis sp.	16	11,85
Bacterial contamination	71	52,59
Epidermophyton sp.	01	0,74
Exserohilum sp.	02	1,48
Fonsecaea sp.	45	33,33
Geotrichum sp.	02	1,48
Microsporum sp.	04	2,96
Penicillium spp.	18	13,33
Pseudallescheria sp.	01	0,74
Scedosporium sp.	01	0,74
Sporothrix sp.	01	0,74
Stacylidium sp.	01	0,74
Trichophyton spp.	12	8,89

Source: Author, 2014.

FIGURE 1 - Isolated colonies of filamentous fungi and yeasts in milk samples from community expansion tanks in the state of Alagoas in 2013.

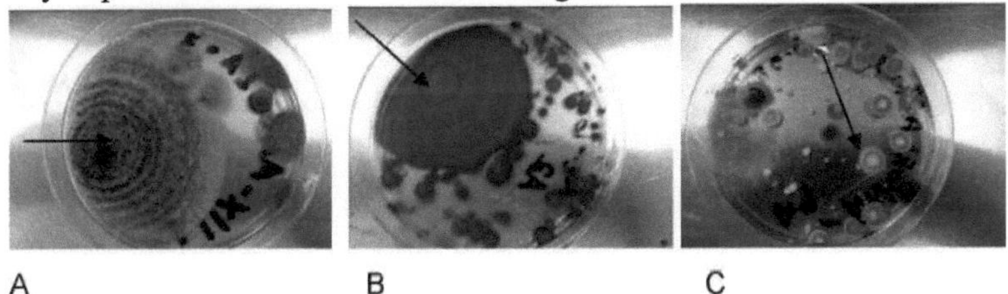

Source: Author, 2014.

A - *Aspergillus* sp. B - *Fonsecaea sp.* C - *Penicillium* sp.

Figure2 Microscopy of isolated colonies of filamentous fungi and yeasts

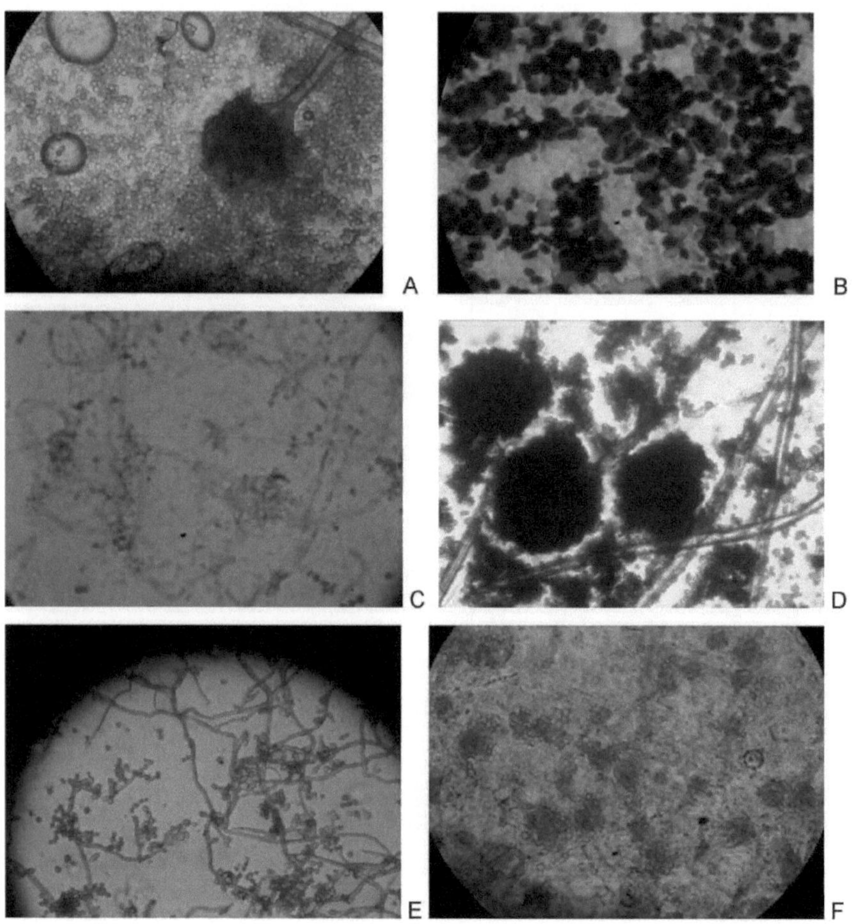

Source: Author, 2014.

A - *Aspergillus* sp

B - *Candida* sp

C - *Penicillium* sp

D - *Aspergillus* sp

E - *Fonsecaea* sp

F - *Penicillium* sp

In this study, a variety of filamentous fungi and yeasts were isolated from all the milk samples collected from community expansion tanks. These microorganisms may be associated with opportunistic mycoses, which are infections caused by low-virulence fungi that can coexist peacefully with the host. However, when conditions favor their development, they can be pathogenic to humans (RUZ-PERES et al., 2010).

VACHEYROU (2011) analyzed raw milk samples from 16 farms in France and found fungi from various genera such as: *Alternaria* spp., other fungi, *Cladosporium spp.*, *Eurotium* spp., white yeasts, red yeasts, *Mucorales spp.*, *Penicillium spp.* and *Wallemia sebi*.

PEREIRA et al. (2013), when observing mesophilic aerobes in UHT milk, found 26.2% of filamentous fungi and yeasts in all the samples. Among the 26.2% of fungi, there was a predominance of filamentous fungi with 67.8% and yeasts accounted for 32.2%. According to the literature, *Candida* sp. is the most common genus present in milk and does not cause significant alterations to the product (TRONCO, 2008). The presence of filamentous fungi and yeasts is indicative of unsatisfactory sanitary practices and, for processed products, may also be related to failures in heat treatment, post-processing contamination or problems with packaging contamination (PEREIRA et al., 2013).

RUZ-PERES et al. (2010), evaluating the resistance of filamentous fungi and yeasts isolated from raw bovine milk to pasteurization and boiling, observed in their study that 27 strains of filamentous fungi; 33.33%, 0% and 0%, respectively, were resistant to rapid pasteurization, slow pasteurization and boiling. With regard to yeast strains, out of a total of 275 strains, 76%, 1.09% and 17.45% were resistant to rapid pasteurization, slow pasteurization and boiling, respectively. Given that the majority of producers of dairy products use rapid pasteurization in their manufacturing technology as a tool for reducing the microbial load of milk, the resistance of yeasts and filamentous fungi after heat treatment represents a risk for the consumption of milk and milk products.

When analyzing the level of fungal contamination in 60 samples of raw milk and 40 samples of dairy products from small industries; TORKAR and VENGUS (2008) observed the presence of fungi in 95% of the raw milk and 63.3% of the dairy products, where 51.5% of the samples were contaminated with the fungus of the genus *Geotrichum*.

DELAVENNE et al. (2011) characterized the fungal contaminants in cow's, goat's and sheep's milk from nine milk samples from tanks located in three different areas in France and found average filamentous fungi and yeast count values ranging from 3.0×10^3 to 4.7×10^4 CFU/mL, regardless of their origin.

MELVILLE (2006) collected milk samples from 50 refrigeration tanks from dairy farms in 3 regions in the interior of the state of São Paulo. Considering the 50 milk samples from refrigeration tanks, the median U.F.C. count of fungi/mL was 28.5×10^2 /mL (0.16 to 332.50 $\times 10^2$ /mL) of milk. The presence of fungi was observed in all milk samples from refrigeration tanks and the following filamentous fungi and yeasts were isolated: *Aureobasidium* spp., *Candida* spp. *(C. krusei, C. parapsilosis, C. kefyr, C. albicans, C. guilliermondii, C. lusitaniae, C. tropicalis), Rhodotorula spp., Geotrichum spp., Trichosporon* spp, *Mucor* spp., *Aspergillus spp., Chrysosporium spp., Acremonium spp., Penicillium spp.* The frequencies of isolations of *C. krusei* (70%), C. *tropicalis* (54%), C. *guilliermondii* (52%), C. *kefyr (34%), Geotrichum spp. (52%)* and *Rhodotorula spp. (40%)*, were statistically higher (p < 0.05) when compared to the other fungi isolated. According to the author, the high count and variety of fungal contamination in milk from expansion tanks could be associated with deficiencies in the hygienic conditions in which the product was obtained and packaged. In the present study, with the exception of the 4.44% of samples in which no colonies grew, the colonies that grew in the rest of the samples were above 5000 CFU of fungi per mL of milk. Among the studies on mold and yeast counts in milk discussed in this paper, it can be seen that this one showed higher contamination than the studies by the authors mentioned above.

Brazilian legislation does not regulate acceptable limits for filamentous fungi and yeasts in milk. However, it has been observed that high counts of these microorganisms indicate inadequate production conditions, as well as unsatisfactory hygiene in the expansion tanks where the milk is stored until it is processed. It should be carefully noted that high counts of these fungi lead to problems such as infections, opportunistic mycoses and even the formation of their metabolites, such as mycotoxins.

Good agricultural, transportation and storage practices remain the best way to prevent food contamination by high counts of filamentous fungi and yeasts. Therefore, strategies and legal procedures are necessary in the farming and food industry to ensure the quality of products of animal origin (QUEIROZ et al., 2009).

The genus *Candida, whose* main species is *C. albicans,* can be isolated from vaginal mucosa, skin, oropharynx and feces, and can cause infections. When characterizing the fungi in this study, it was observed that 14.07% of the samples had yeasts of the genus *Candida.* ZARAGOZA et al. (2011), in their study to identify yeasts in the mammary glands of animals with mastitis, analyzed the milk of healthy animals and those at different stages of mastitis. In his research, 44.32% of the milk from the animals analyzed had clinical mastitis. She also observed that 25.75% of the total milk analyzed had yeasts of the genus *Candida of* different species.

In order to obtain good quality milk, care must be taken primarily in the field, where milk is most subject to contamination. This requires the adoption of good agricultural practices and the hygienic production of milk. Thus, milkers should receive regular training in order to minimize all forms of contamination in the milk. It is of fundamental importance to cool the milk using expansion tanks; however, this cooling should be done as quickly as possible and should take place immediately after milking, in previously cleaned and sanitized tanks (TEIXEIRA, 2014). In this study, a considerable number of fungal genera were observed in the milk samples from the community expansion tanks, and only by using appropriate hygienic measures will it be possible to reduce the microbial load and consequently avoid the problems caused by these microorganisms.

4. CONCLUSION

Fungal contamination in milk stored in expansion tanks is due to the following factors: poor milking hygiene, inadequate storage temperature, heterogeneous samples from different farms, inadequate facilities in expansion tanks; these factors contribute to an increase in the microbial load in milk, making it a risk to public health.

The most contaminated fungal microorganisms in milk from community expansion tanks in the state of Alagoas are filamentous, and the high counts and wide variety of these contaminants probably derive from inadequate farming practices and insufficient hygiene in the expansion tanks.

REFERENCES

AMERICAN PUBLIC HEALTH ASSOCIATION (APHA). Compendium of methods for the microbiological examination of foods. 4. ed. Washington: American Public Heath Association. p. 663-667, 2001.

COSTA, G.M., SILVA, N.; ROSA, C. A. Yeast mastitis in dairy cattle in southern Minas Gerais State, Brazil. **Ciência Rural**, v.38, p. 1938-1942, 2008.

DELAVENNE E. et al. Fungal diversity in cow, goat and ewe milk. **International Journal of Food Microbiology**. v.151, p. 247-251,2011.

LACAZ, C.S. et al. Treatise on medical mycology. 9.ed. São Paulo: Sarvier, p. 15829, 2002.

MELVILLE, P.A. Occurrence of fungi in raw milk from refrigeration tanks and cans on dairy farms, as well as milk sold directly to consumers. Arquivo Instituto Biológico, São Paulo, v.73, n.3, p.295-301, jul./set., 2006.

PEREIRA, et. al. Microbiota mesophilic aerobic contaminant UHT milk. **Revista Instituto Laticínios Cândido Tostes**, Juiz de Fora, v. 68, n°. 394, p. 25-31, 2013.

QUEIROZ, V. A. V. et al. **Good practices and HACCP system in the post-harvest phase of corn**. Sete Lagoas: Embrapa Milho e Sorgo. Technical circular, no.122. p.28, 2009.

RIBEIRO, M.G. Therapeutic principles in mastitis in production and companion animals. In: ANDRADE S.F. (Ed.). **Manual of veterinary therapeutics.** 3. ed. São Paulo: Roca, p. 759-771, 2008.

RUZ-PERES M. et al. Resistance of filamentous fungi and yeasts isolated from raw bovine milk to pasteurization and boiling. **Arquivo Brasileiro de Medicina**

Veterinária e Zootecnia. v.17, n°.1, p.62-70, 2010.

SILVA, N. et. al. **Manual de métodos de análise microbiológica de alimentos.** contagem de fungos filamentosos e leveduras. 3. ed. São Paulo: Varela, 2007. Chap. 2. p. 99-108.

SPANAMBERG A. et. al. Mycotic mastitis in ruminants caused by yeasts. **Ciência. Rural.** v. 39 n°.1, p. 20, 2008.

TEIXEIRA, S. Milk cooling tanks: a solution for dairy farmers. Available at: http://www.cpt.com.br/cursos-bovinos-gadodeleite/artigos/tanques-de-resfriamento-de-leite-uma-solucao-para-os- dairy-producers. accessed on: June 2014.

TORKAR, K.G & VENGUS, T. The presence of yeasts, moulds and aflatoxin M1 in raw Milk and cheese in Slovenia. **Food Control**, v.19, p.570-577, 2008.

TRONCO, V.M. **Manual for milk quality inspection**. 3ed. Santa Maria: UFSM, 2008. 206p.

VACHEYROU M. et al. Cultivable microbial communities in raw cow milk and potential transfers from stables of sixteen French farms. **International Journal of Food Microbiology**. 146, p. 253-262, 2011.

ZARAGOZA C. S. al. Yeasts isolation from bovine mammary glands under different mastitis status in the Mexican High Plateu. **Revista Iberoamericana de Micología**. México . v.28, n°2, p.79-82, 2011.

I want morebooks!

Buy your books fast and straightforward online - at one of world's fastest growing online book stores! Environmentally sound due to Print-on-Demand technologies.

Buy your books online at
www.morebooks.shop

Kaufen Sie Ihre Bücher schnell und unkompliziert online – auf einer der am schnellsten wachsenden Buchhandelsplattformen weltweit! Dank Print-On-Demand umwelt- und ressourcenschonend produziert.

Bücher schneller online kaufen
www.morebooks.shop

info@omniscriptum.com
www.omniscriptum.com

Printed by Books on Demand GmbH, Norderstedt / Germany